Finding the Light

How Life Always Finds a Way

Authored by
Zahid Ameer

Published by

Goodword eBooks

Copyright © 2024 Zahid Ameer

All rights reserved.

ISBN: 9798340716323

DEDICATION

"I dedicate this book to my beloved parents, whose wisdom I hold in the highest regard. Their every word of guidance has been a beacon of light, illuminating the path of my life and shaping the essence of who I am."

Finding the Light

Contents:

Contents:

Introduction: The Resilience of Life

Chapter 1: The Origins of Life and Survival

Chapter 2: Adaptation and Evolution: Nature's Master Plan

Chapter 3: Life in Extreme Environments

Chapter 4: The Struggle for Existence in the Animal Kingdom

Chapter 5: The Human Spirit: Overcoming Adversity

Chapter 6: Triumph of the Will: Stories of Survival

Chapter 7: Nature's Recovery After Disasters

Chapter 8: Microbial Resilience – The Hidden World of Survival

Chapter 9: Healing and Rebirth in Ecosystems

Chapter 10: Finding Hope in Life's Darkest Moments

Conclusion: Life's Eternal Quest for the Light

Bibliography

Finding the Light

Acknowledgments

Disclaimer:

About me

Finding the Light

Introduction: The Resilience of Life

Life is a force of nature, a phenomenon that defies the odds and persists in the most unlikely of circumstances. From the dawn of Earth's history, life has continually demonstrated an extraordinary ability to evolve, adapt, and survive. The resilience of life is woven into the fabric of our planet's story, revealing itself in ways both grand and minute, in organisms large and small, in environments familiar and foreign. Whether in the crushing pressures of the ocean's abyss, the desolate icy expanses of polar regions, or the scalding heat of volcanic landscapes, life finds a way to endure.

The Origins of Life's Resilience

The earliest life forms likely emerged around 3.5 to 4 billion years ago, a time when Earth was a hostile, turbulent place. The atmosphere was filled with toxic gases, temperatures swung between extremes, and the planet was bombarded by meteorites. And yet, life found a way to emerge from these harsh conditions. Scientists believe that the first life forms were simple, single-celled organisms, perhaps arising from hydrothermal vents deep in the ocean or shallow pools rich in organic molecules. These early organisms had to contend with a volatile environment, but it was their ability to adapt to changing conditions that laid the foundation for life's persistence on Earth.

From these humble beginnings, life diversified into an astonishing array of forms. The process of natural selection, as described by Charles Darwin, provided life with a powerful tool for resilience. Organisms that were better suited to their environments survived and reproduced, while others perished. This mechanism allowed life to not only persist but to thrive, constantly evolving in response to environmental pressures, finding new ways to exploit resources, and developing ingenious strategies for survival.

Life in Extreme Environments: A Testament to Resilience

One of the most profound demonstrations of life's resilience is its ability to thrive in extreme environments, known as extreme ecosystems. These ecosystems challenge the very definition of what we consider "habitable," yet life flourishes in them. Extremophiles—organisms that live in conditions once thought too severe to support life—are some of the most remarkable examples of life's adaptability.

1. **Deep Ocean Hydrothermal Vents:** At the bottom of the ocean, where sunlight never penetrates,

Finding the Light

hydrothermal vents spew superheated water rich with chemicals. Despite the darkness and crushing pressure, entire ecosystems exist around these vents, supported not by photosynthesis, but by chemosynthesis—a process where bacteria convert chemicals like hydrogen sulfide into energy. These bacteria form the foundation of a food chain that includes giant tube worms, clams, and shrimp, all thriving in an environment previously thought to be inhospitable to life.

2. **Polar Extremes:** In the frozen tundras of the Arctic and Antarctic, temperatures plunge far below freezing, and food is scarce. Yet, life persists. Microorganisms, plants, and animals have developed incredible adaptations to survive these extreme conditions. Antarctic fish, for example, produce "antifreeze" proteins that prevent their blood from freezing. Polar bears and penguins have evolved thick layers of fat and fur to insulate themselves against the cold, while migratory birds and animals have learned to time their breeding cycles with the brief summer months, maximizing their chances of survival.

3. **Desert Life:** In some of the hottest and driest places on Earth, such as the Sahara or the Atacama Desert, life seems almost impossible. Yet, plants and animals have evolved remarkable ways to cope with

Finding the Light

extreme heat and lack of water. Cacti store water in their thick stems, while some desert animals, like the kangaroo rat, can survive without drinking water by obtaining moisture from the seeds they eat and conserving water through highly efficient kidneys.

These extreme environments are not anomalies, but rather testaments to life's adaptability. They show that life is not only capable of surviving under favorable conditions but can flourish in places that seem entirely uninhabitable.

The Struggle for Survival: Adaptation and Evolution

The resilience of life is not limited to thriving in extreme conditions. It is equally evident in the ongoing struggle for survival within more typical ecosystems. Every living organism must contend with a host of challenges, from finding food and water to avoiding predators and disease. Yet, life's ingenuity shines through in the myriad ways species have evolved to meet these challenges.

- **Predator-Prey Dynamics:** In the animal kingdom, survival is often a matter of life and death. Predators must constantly hone their hunting skills to capture their prey, while prey species evolve new defenses to avoid being eaten. Camouflage, speed, strength, and agility are just a few of the adaptations that allow species to survive in this constant battle. For instance, the gazelle's agility and speed have evolved in response to the lion's pursuit, creating a dynamic balance in which both predator and prey must continually adapt.

- **Reproductive Strategies:** Life's resilience is also evident in the variety of reproductive strategies species employ to ensure the survival of their genes. Some species produce vast numbers of offspring to ensure that at least a few survive to adulthood, while others invest significant resources into raising a smaller number of young. This diversity of strategies, from the mass spawning of fish to the nurturing care of mammals, demonstrates life's ability to adapt to different environments and challenges.
- **Migration and Seasonal Adaptations:** Many species have developed migratory behaviors that allow them to exploit different environments at different times of the year. Birds, for example, travel thousands of miles to breed in favorable conditions, while certain fish, such as salmon, migrate between freshwater and saltwater environments to complete their life cycles. These migrations are often perilous, yet they are vital for survival, showcasing the lengths to which life will go to persist.

Life After Catastrophe: The Ultimate Test of Resilience

Finding the Light

Perhaps one of the most awe-inspiring aspects of life's resilience is its ability to recover after catastrophic events. Throughout Earth's history, life has endured mass extinctions, volcanic eruptions, asteroid impacts, and climate shifts that have wiped out vast numbers of species. Yet, in the aftermath of these disasters, life has always rebounded, often diversifying into new forms and filling ecological niches left vacant by extinction.

1. **The Permian-Triassic Extinction:** Known as the "Great Dying," this event occurred around 252 million years ago and wiped out approximately 96% of marine species and 70% of terrestrial species. Despite this near-total annihilation of life, the Earth's ecosystems eventually recovered, giving rise to the age of the dinosaurs.
2. **The Cretaceous-Paleogene Extinction:** 66 million years ago, an asteroid impact led to the extinction of the dinosaurs and many other species. In the aftermath, mammals—previously small and insignificant—began to thrive, eventually leading to the rise of humans.

These events illustrate that while individual species may perish, life as a whole is incredibly resilient. It adapts, evolves, and rebounds in ways that are often unpredictable, filling the gaps left by extinction with new and diverse forms of life.

Finding the Light

Life's Unyielding Persistence in the Face of Uncertainty

The resilience of life is not only a biological phenomenon but a philosophical one as well. Life, in all its forms, seems to possess an unyielding will to persist. Whether it's a plant sprouting from a crack in the pavement, a bird migrating across thousands of miles, or a human overcoming personal tragedy, life's determination to find the light, to endure, and to flourish is a common thread that unites all living beings.

In the chapters that follow, we will explore the many facets of life's resilience. From the origins of life and the evolutionary processes that drive adaptation to the incredible survival strategies employed by species in extreme environments, this book will delve into the mechanisms that allow life to thrive. We will also explore the human dimension of resilience, examining how individuals and communities overcome adversity and find hope in even the darkest moments.

Ultimately, the story of life is one of triumph against the odds. It is a story of light emerging from darkness, of life finding a way where none seems possible. Through these pages, we will celebrate life's incredible capacity to endure, adapt, and evolve, reaffirming the age-old truth that, no matter how challenging the circumstances, life always finds a way.

Chapter 1: The Origins of Life and Survival

The origin of life on Earth is a story of resilience, adaptation, and persistence, beginning at the molecular level over 3.5 billion years ago. Despite catastrophic events that have threatened the existence of all life forms, including asteroid impacts, volcanic eruptions, and mass extinctions, life has endured. It began in a world radically different from our own, with simple organisms that developed survival strategies enabling them to persist through the most extreme conditions. This chapter explores the conditions under which life began, the earliest forms of life, and how mass extinctions shaped the resilience and evolution of living organisms.

The Early Earth: A Hostile Beginning

The Earth formed approximately 4.5 billion years ago, emerging from a cloud of gas and dust orbiting the young Sun. The early Earth was a harsh and violent environment, unrecognizable compared to the planet we know today. Volcanic activity was rampant, with massive eruptions spewing lava and gases into the atmosphere. The Earth's surface was bombarded by asteroids and comets, and the planet was still cooling from its initial molten state.

1. The Atmosphere

The early atmosphere was devoid of oxygen, which is essential for the life forms we are familiar with today. Instead, it was composed primarily of carbon dioxide, methane, ammonia, water vapor, and nitrogen. This anoxic environment (lacking oxygen) created challenges for the emergence of life but also provided the necessary building blocks for the first biochemical reactions.

The absence of an ozone layer allowed harmful ultraviolet (UV) radiation to penetrate the atmosphere. The surface of the planet was also subjected to extreme temperature fluctuations. Despite these seemingly inhospitable conditions, the basic chemistry needed to create life was occurring, possibly sparked by lightning storms or volcanic activity.

2. The Oceans: Cradle of Life

The formation of Earth's oceans was a pivotal moment in the development of life. As the planet cooled, water vapor in the atmosphere condensed and fell as rain, forming vast oceans. These oceans became the "primordial soup," a nutrient-rich environment where simple molecules could combine to form more complex structures. The water in the

Finding the Light

oceans provided protection from UV radiation and helped to stabilize the chemical reactions that would lead to the formation of the first life forms.

Deep-sea hydrothermal vents may have played a crucial role in the origin of life. These vents, located at the bottom of the ocean, spew mineral-rich water heated by the Earth's interior. These minerals provided the necessary chemicals for life's earliest biochemical processes. Some scientists propose that these deep-sea environments, with their stable temperatures and abundance of chemical energy, were ideal locations for the origin of life.

First Signs of Life: The Rise of Simple Organisms

The first life forms to emerge were simple, single-celled organisms. These organisms were prokaryotes—life forms without a defined nucleus. Prokaryotes include bacteria and archaea, both of which are thought to have evolved early in Earth's history. These microorganisms were the pioneers of life, and they developed remarkable strategies to survive in extreme and volatile environments.

1. The First Self-Replicating Molecules

Life, at its core, is defined by the ability to reproduce and pass genetic information from one generation to the next.

Scientists believe that life began with the formation of simple self-replicating molecules, such as RNA (ribonucleic acid). RNA is capable of both storing genetic information and catalyzing chemical reactions, making it a likely candidate for the first molecule that could carry out life-like functions. The emergence of RNA or a similar molecule marked the beginning of biological evolution.

Over time, these self-replicating molecules became encapsulated within membranes, forming the first cells. These protocells were likely simple, composed of fatty acids forming a boundary that enclosed genetic material and the biochemical machinery needed for replication.

2. Bacteria and Archaea: The Early Survivors

The earliest forms of life on Earth were likely similar to modern bacteria and archaea. These microorganisms thrived in environments that were inhospitable to most modern life forms. Archaea, in particular, are known for their ability to survive in extreme environments, including hot springs, acidic waters, and highly saline environments.

- **Bacteria** are incredibly versatile and were among the first life forms to harness energy through processes such as fermentation and photosynthesis. Some of the earliest bacteria were likely anaerobic, meaning they did not require oxygen for survival. Instead, they relied on chemical reactions that did

not involve oxygen, such as sulfur or hydrogen-based metabolism.
- **Archaea**, though similar to bacteria in appearance, are genetically distinct and possess unique biochemical pathways that allow them to survive in extreme conditions, including temperatures over 100°C (212°F) in deep-sea hydrothermal vents or highly acidic environments.

3. Photosynthesis: The Oxygen Revolution

One of the most significant evolutionary developments in Earth's history was the evolution of photosynthesis. Cyanobacteria, a group of bacteria capable of photosynthesis, began to produce oxygen as a byproduct of converting sunlight into chemical energy. This process, which began around 2.5 billion years ago, fundamentally altered the Earth's atmosphere.

The "Great Oxygenation Event" occurred when oxygen levels in the atmosphere began to rise, marking the transition to an oxygen-rich environment. This oxygen revolution was a double-edged sword: while it allowed for the evolution of more complex life forms, it also led to the extinction of many anaerobic organisms that were unable to tolerate oxygen.

Mass Extinctions: Life's Resilience and Rebirth

Throughout Earth's history, life has been repeatedly tested by mass extinction events—catastrophic occurrences that wiped out vast numbers of species in a relatively short period. These events were often caused by natural disasters, such as asteroid impacts, volcanic eruptions, and drastic climate changes. While mass extinctions caused the demise of many species, they also played a crucial role in shaping the resilience and adaptability of life.

1. The Role of Mass Extinctions

Mass extinctions are nature's way of reshuffling the biological deck. They create opportunities for the surviving species to evolve and fill new ecological niches. Following each extinction event, the evolutionary landscape changes dramatically, allowing new forms of life to emerge and diversify.

One of the earliest and most significant extinction events was the **Permian-Triassic extinction**, which occurred around 252 million years ago and wiped out approximately 96% of marine species and 70% of terrestrial vertebrates. This event was likely triggered by massive volcanic eruptions, which led to severe global warming, ocean acidification, and the release of toxic gases into the atmosphere. Despite the magnitude of this extinction, life

Finding the Light

eventually rebounded, leading to the rise of the dinosaurs and other complex life forms.

2. The Cretaceous-Paleogene Extinction

The most famous mass extinction event occurred 66 million years ago at the end of the Cretaceous period. This event, caused by an asteroid impact near present-day Mexico, led to the extinction of the dinosaurs and many other species. However, this disaster also paved the way for the rise of mammals, which had previously lived in the shadow of the dinosaurs. The extinction of dominant species creates evolutionary space for others to flourish, and in this case, mammals took advantage of the opportunity, eventually leading to the evolution of humans.

3. Survival Through Catastrophe

What makes life so resilient in the face of mass extinctions is its ability to adapt and evolve. Organisms that can rapidly adjust to new environmental conditions are more likely to survive. After each extinction event, species that survived were typically those with versatile diets, adaptable reproductive strategies, or the ability to thrive in multiple environments.

Mass extinctions highlight an important truth about life: while individual species may perish, life itself persists and adapts. From the ashes of extinction, new forms of life arise, often more suited to the changing conditions of the Earth.

Conclusion: Life's Persistence Against All Odds

The origins of life on Earth reveal a remarkable story of survival and adaptation in the face of adversity. From the emergence of simple self-replicating molecules in Earth's primordial oceans to the evolution of resilient bacteria and archaea, life has continuously found a way to persist in the harshest of conditions. Mass extinctions, though devastating, have played a critical role in shaping the trajectory of evolution, allowing life to renew and diversify. As we move forward in this exploration of life's resilience, we will uncover even more examples of how life, in its myriad forms, always finds a way to survive and thrive.

Chapter 2: Adaptation and Evolution: Nature's Master Plan

Life's resilience is not a coincidence; it's the result of millions of years of evolution, a process that has shaped every living organism on Earth. At the heart of this process is **natural selection**, the concept introduced by Charles Darwin in his groundbreaking work, *On the Origin of Species* (1859). Darwin's theory explains how species evolve and adapt to their environments over time, resulting in a dynamic and ever-changing natural world where survival depends on an organism's ability to respond to environmental pressures.

Natural selection acts as the engine of evolution, continuously refining species so that they are better equipped to survive and reproduce in their environments. Over generations, the traits that enhance survival become more common in the population, while less favorable traits are gradually eliminated. This process of gradual change allows life to "find the light" even in the harshest conditions, ensuring that those who can adapt, survive, and thrive.

Darwin's Theory of Evolution: The Mechanisms of Natural Selection

Charles Darwin's theory of evolution by natural selection was revolutionary for its time, offering an explanation for the diversity of life on Earth that didn't rely on a static or unchanging world. Darwin proposed that organisms vary in their traits, and these variations can be inherited by their offspring. However, not all organisms survive to reproduce; those with traits better suited to their environment are more likely to survive and pass on those traits. Over long periods, this process leads to the evolution of species.

The core components of natural selection are:

1. **Variation**: Within a population, individuals exhibit variation in their traits. These differences could be in size, color, speed, or any number of physical or behavioral characteristics. These variations often result from mutations in the genetic code, sexual reproduction, and genetic recombination.
2. **Heritability**: Many of these variations are heritable, meaning they can be passed from parents to their offspring. This is crucial because natural selection operates on traits that can be inherited, ensuring that beneficial traits persist across generations.

3. **Differential Survival and Reproduction**: In any environment, organisms face competition for resources such as food, shelter, and mates. Those that possess traits better suited to the environment are more likely to survive and reproduce. For example, faster animals may escape predators more easily, while plants that can store water effectively may survive in drought conditions.
4. **Accumulation of Favorable Traits**: Over many generations, advantageous traits accumulate within a population, leading to a gradual evolution of the species. The accumulation of these traits results in organisms that are more adapted to their environment.

Darwin famously coined the phrase **"survival of the fittest"**, though it was Herbert Spencer who first used the term in describing Darwin's work. "Fittest" here refers not necessarily to strength or power, but to an organism's fitness, or its ability to survive and reproduce in a given environment. The organisms that can best "fit" into their environment, through traits that give them an edge, are more likely to pass on their genes, and over time, these traits become more prevalent.

Adaptation: How Species Change Over Time to Suit Their Environment

Finding the Light

Adaptation is the key outcome of natural selection. An adaptation is any characteristic—physical, behavioral, or physiological—that enhances an organism's ability to survive and reproduce in a specific environment. Over long periods, these adaptations can lead to the evolution of new species, a process known as **speciation**.

Adaptations can be categorized into three types:

1. **Structural Adaptations**: These are physical features of an organism that enhance survival. Examples include the long neck of a giraffe, which allows it to reach food high in trees, or the thick fur of Arctic animals, which helps them retain heat in freezing climates.
2. **Behavioral Adaptations**: These involve changes in how an organism behaves to better suit its environment. For example, many birds migrate to warmer regions during winter to access food, and nocturnal animals are active at night to avoid predators and extreme temperatures.
3. **Physiological Adaptations**: These are internal processes that improve an organism's survival chances. For instance, desert animals often have efficient water-conservation mechanisms that allow them to survive with minimal water. Some fish, like salmon, can regulate their internal salt concentration,

enabling them to move between freshwater and saltwater environments.

Examples of Evolution in Action

The theory of natural selection can be observed in the natural world through numerous examples where organisms have adapted to environmental pressures over time. Two of the most famous examples of natural selection in action are the **Galápagos finches** and the **peppered moths** during the Industrial Revolution.

The Galápagos Finches

One of the most iconic examples of natural selection involves the **finches of the Galápagos Islands**, often referred to as "Darwin's finches." During his voyage on the HMS Beagle, Darwin observed that finches on different islands had different beak shapes and sizes, despite being closely related. These variations were adaptations to the specific environments and food sources available on each island.

- **Large Beaks for Hard Seeds**: On islands where finches primarily ate large, hard seeds, birds with larger, stronger beaks were better able to crack open the seeds. Over time, the finch population on these islands developed larger beaks as individuals with this trait were more likely to survive and reproduce.

- **Thin Beaks for Insects**: In contrast, on islands where insects were the primary food source, finches with thinner, more precise beaks thrived, as they could better catch insects. Natural selection favored those with thinner beaks, leading to a predominance of that trait in the population.

The diversity in beak shape and size among the finch species demonstrated how species could adapt to their environment through gradual changes over time. This process, known as **adaptive radiation**, occurs when a single species evolves into several species, each adapted to different niches.

The Peppered Moth: A Classic Example of Natural Selection

Another well-documented case of natural selection involves the **peppered moth (Biston betularia)** during the Industrial Revolution in England. Before industrialization, most peppered moths were light-colored, which allowed them to blend in with the lichen-covered trees in their environment, helping them avoid predators like birds. However, as industrial pollution increased, the soot from factories darkened the trees by killing the lichen.

- **Industrial Melanism**: As the trees darkened, the lighter-colored moths became more visible to predators, while darker-colored (melanic) moths,

Finding the Light

which were previously rare, blended in better with the soot-covered trees. Over time, the population of dark-colored moths increased because they were more likely to survive and reproduce, a phenomenon known as **industrial melanism**.

- **Reversal After Clean Air Acts**: After environmental regulations were introduced to reduce pollution, the trees began to lighten again as lichen returned, and the population of light-colored moths rebounded, demonstrating natural selection's dynamic nature.

This rapid evolutionary shift in the moth population provided direct evidence of how environmental changes can drive natural selection and lead to observable changes in species over a relatively short period.

Conclusion: The Power of Evolution in Shaping Life

Evolution, driven by natural selection, is nature's most powerful mechanism for ensuring the survival of life. Through countless generations, species have adapted to an ever-changing world, finding the light in darkness by developing traits that enhance their survival and reproductive success. Whether it's the unique beaks of Galápagos finches, the coloration of peppered moths, or the astonishing adaptations of extremophiles, the process of evolution ensures that life can persist, even in the most

Finding the Light

hostile environments. Darwin's theory of natural selection not only provides a profound understanding of life's diversity but also explains how life's journey continues, forever adapting to find its way forward.

Chapter 3: Life in Extreme Environments

Life's extraordinary resilience is most evident in extremophiles, organisms that thrive in environments so harsh that they were once believed to be completely inhospitable to life. These organisms defy what were once considered the essential prerequisites for life, pushing the limits of biological endurance and expanding our understanding of how adaptable life can be. Extremophiles are not only fascinating because of their resilience, but they also provide insights into how life might exist in extreme conditions elsewhere in the universe, such as on Mars or the icy moons of Jupiter and Saturn.

In this chapter, we'll explore how life survives in some of the most extreme environments on Earth, from the crushing depths of the ocean floor, to the scorching deserts, and the freezing cold of the polar regions. We'll also look at how organisms have evolved incredible adaptations to live in these environments, demonstrating that life, in its endless forms, truly does find a way.

Deep Sea Vents: Life in the Absence of Sunlight

One of the most remarkable examples of life thriving in extreme environments is the discovery of ecosystems

Finding the Light

living around hydrothermal vents on the ocean floor. These vents, first discovered in the 1970s, are cracks in the Earth's surface where seawater, heated by molten rock beneath the Earth's crust, gushes out at temperatures that can exceed 400°C (752°F). What makes these environments so unique is that they are completely devoid of sunlight, the energy source that powers most life on Earth through photosynthesis.

Yet, despite the lack of sunlight, entire ecosystems thrive around these vents. The key to life here is a process called **chemosynthesis**. Instead of relying on sunlight for energy, bacteria around the hydrothermal vents use chemicals like hydrogen sulfide, which is abundant in the vent emissions, to produce energy. These bacteria form the foundation of the food chain in this extreme environment, much like plants do on the surface.

- **Tube Worms:** One of the most iconic creatures living near hydrothermal vents is the giant tube worm (*Riftia pachyptila*). These worms, which can grow over 2 meters (6.5 feet) in length, have no digestive system. Instead, they house symbiotic bacteria in a specialized organ called a trophosome. These bacteria use chemosynthesis to convert sulfur compounds into energy, which the worms then use to survive. In return, the worms provide the bacteria with the chemicals they need to thrive.

- **Pompeii Worms:** Another fascinating creature found near hydrothermal vents is the Pompeii worm (*Alvinella pompejana*). This worm lives in temperatures that can exceed 80°C (176°F), making it one of the most heat-tolerant animals on the planet. The worm's back is covered in a thick layer of bacteria, which may help it survive in these extreme temperatures by providing a protective layer against the heat.
- **Vent Crabs and Fish:** Larger animals, such as crabs, shrimp, and even fish, also make their home near hydrothermal vents. These creatures feed on the bacteria or smaller animals that have adapted to this extreme environment. Despite the seemingly harsh conditions, these ecosystems are surprisingly diverse, demonstrating that life can flourish even in the most unexpected places.

The discovery of these hydrothermal vent ecosystems has revolutionized our understanding of where life can exist. It has also fueled speculation about the potential for life on other celestial bodies, such as Jupiter's moon Europa, which is thought to have a subsurface ocean where similar hydrothermal activity might occur.

Desert Life: Surviving with Minimal Water

Finding the Light

Deserts represent some of the most extreme environments on Earth due to their extreme temperatures and lack of water. Yet, life has found incredible ways to adapt to these harsh conditions. From the vast, scorching sands of the Sahara to the bone-dry Atacama Desert in South America, both plants and animals have developed remarkable strategies to survive with minimal water.

- **Desert Plants:** One of the most famous desert survivors is the **cactus**. Cacti have evolved to store water in their thick, fleshy tissues, allowing them to survive long periods of drought. Their spines, which are modified leaves, not only protect them from herbivores but also help reduce water loss by providing shade and reducing airflow around the plant. The **Saguaro cactus**, for example, can store hundreds of gallons of water during a single rainfall, which it slowly uses during dry periods.
  Another adaptation seen in desert plants is **CAM photosynthesis** (Crassulacean Acid Metabolism). In this process, plants open their stomata (pores) at night, rather than during the day, to take in carbon dioxide. This reduces water loss, as cooler nighttime temperatures slow down the evaporation of water.
- **Desert Animals:** Many desert animals are nocturnal, avoiding the extreme heat of the day by becoming active only at night. The **kangaroo rat**, for instance, gets almost all of its water from the seeds it eats and

rarely, if ever, drinks water. It can survive in some of the driest deserts by producing very concentrated urine and dry feces, conserving as much water as possible.

The **Fennec fox**, with its large ears, helps regulate its body temperature by dissipating heat. Its light-colored fur also reflects sunlight, helping it stay cool during the day, while its burrowing behavior provides a cool refuge from the harsh desert sun.

Camels, often called the "ships of the desert," are perhaps the most iconic desert survivors. Contrary to popular belief, their humps do not store water, but fat, which they can metabolize when food is scarce. They also have an extraordinary ability to survive without water for long periods and can drink up to 40 gallons of water in one go when they do find it. Their nostrils can close to keep out blowing sand, and their thick coats protect them from both extreme heat during the day and cold temperatures at night.

Polar Extremes: Life at the Ends of the Earth

At the opposite end of the spectrum from deserts are the polar regions, where life must survive in extreme cold, limited food availability, and long periods of darkness. Both the Arctic and Antarctica are some of the most hostile environments on Earth, yet life has adapted to survive in these frozen wastelands.

- **Microscopic Survivors:** One of the most resilient organisms found in polar regions is the **tardigrade**, also known as the "water bear." These tiny, nearly indestructible creatures can survive extreme temperatures, from the near-absolute zero cold of space to boiling water. In polar regions, tardigrades enter a state called **cryptobiosis**, where they lose almost all of their water and effectively shut down their metabolism. In this state, they can survive freezing temperatures and remain dormant until conditions improve.
- **Antarctic Fish:** The waters surrounding Antarctica are some of the coldest on Earth, yet fish like the **Antarctic toothfish** have evolved antifreeze proteins in their blood, which prevent their bodies from freezing. These proteins bind to ice crystals and inhibit their growth, allowing the fish to swim in waters that would freeze the blood of most other fish species.
- **Polar Bears and Seals:** In the Arctic, **polar bears** are the top predators. Their thick fur and a layer of blubber beneath their skin provide insulation against the freezing temperatures. Polar bears are excellent swimmers, capable of covering long distances in the icy Arctic waters, and their keen sense of smell helps them locate seals, their primary food source, under the ice.

Finding the Light

Seals, like the Weddell seal in Antarctica, have adapted to the extreme cold by developing thick layers of blubber, which not only keep them warm but also serve as an energy reserve during the long winter months when food can be scarce. Seals can also hold their breath for extended periods, allowing them to dive deep beneath the ice to find food.

- **Penguins:** In Antarctica, **emperor penguins** are well adapted to the extreme cold. These birds huddle together in large groups to conserve heat during the harsh Antarctic winter, where temperatures can drop to -60°C (-76°F). Their feathers provide excellent insulation, and a thick layer of blubber beneath their skin helps them survive both in and out of the icy waters. Emperor penguins also fast for months while incubating their eggs, relying on stored body fat for survival.

Conclusion: Life's Endless Adaptability

The existence of life in extreme environments—from the crushing depths of the ocean to the blistering heat of deserts and the freezing cold of the poles—demonstrates the incredible adaptability of living organisms. These extremophiles challenge our understanding of the conditions necessary for life and suggest that life, in its

Finding the Light

endless diversity, may be far more common in the universe than previously thought.

By studying these extraordinary survivors, scientists gain not only a deeper understanding of how life has evolved on Earth but also clues as to where we might find life elsewhere in the cosmos. Whether it's in the icy subsurface oceans of Jupiter's moon Europa or the scorching surface of Venus, life may be more tenacious than we ever imagined.

Chapter 4: The Struggle for Existence in the Animal Kingdom

In the animal kingdom, survival is an ongoing struggle. Every species is engaged in a perpetual contest for food, shelter, and the continuation of their lineage. This constant challenge shapes the way animals behave, evolve, and interact with one another. The balance of predator and prey, the intricate survival strategies that evolve over time, and the measures taken to ensure offspring survival are all crucial components of this struggle. Through millions of years of evolution, animals have developed complex and diverse adaptations, transforming them into specialists within their environments. This chapter delves into three critical aspects of the natural world's survival game: predator-prey relationships, survival strategies, and parental care.

Predator-Prey Relationships: The Balance Between Life and Death

One of the most dramatic and visible manifestations of the struggle for survival in the animal kingdom is the relationship between predators and their prey. This relationship is a delicate balance of life and death, a constant evolutionary arms race where both sides adapt to outwit or outrun the other. Predators, as hunters, are continually evolving to become more efficient in capturing their prey, while prey species are simultaneously developing mechanisms to evade and survive these threats.

The Arms Race: Evolutionary Adaptations

The relationship between predators and prey is often described as an "evolutionary arms race" because both sides are constantly under pressure to adapt. Predators must evolve to become better hunters, developing speed, stealth, and powerful weapons such as claws, sharp teeth, or venom. Meanwhile, prey species must evolve to evade predators, often becoming faster, more agile, or equipped with enhanced senses to detect danger.

For example, consider the cheetah and the gazelle. Cheetahs are renowned for their extraordinary speed, capable of reaching up to 60 miles per hour in short bursts. This speed is a direct evolutionary response to the gazelle's agility and swiftness. Gazelles, in turn, have evolved to be quick and capable of sharp, unpredictable turns, making it difficult for the cheetah to catch them during a chase. This predator-prey dynamic keeps both species locked in a continuous cycle of evolutionary advancement, where the survival of one influences the adaptation of the other.

Stealth and Ambush Tactics

Finding the Light

Some predators rely on stealth and ambush rather than speed. For instance, big cats like lions and tigers often hunt by stalking their prey and getting as close as possible before launching a quick, powerful attack. This strategy minimizes energy expenditure and increases the chances of success. These predators have evolved to have camouflaged coats that help them blend into their surroundings, allowing them to approach their prey unnoticed.

Snakes, particularly those that use venom, also exemplify ambush predators. For example, pit vipers, such as rattlesnakes, use heat-sensitive pits to detect warm-blooded prey in the dark. They strike quickly, delivering venom that immobilizes their prey before consuming it. The venom itself is an evolved weapon that ensures the predator doesn't have to chase or overpower its prey, relying instead on chemical warfare to ensure a successful hunt.

Defensive Adaptations of Prey

In response, prey species have developed a range of defenses. Some animals, such as the porcupine, rely on physical deterrents—sharp quills or spines that make them difficult to attack. Others, like the skunk, employ chemical defenses, spraying foul-smelling liquids to repel potential predators.

A particularly fascinating example is the bombardier beetle, which has evolved to eject a boiling, noxious chemical spray from its abdomen when threatened. The spray is a mixture of hydrogen peroxide and hydroquinone, which, when combined, react violently, creating a hot, stinging blast that can deter even the most persistent predators.

Survival Strategies: Camouflage, Mimicry, Speed, and Strength

The ability to survive in the wild depends not only on an animal's physical traits but also on its behavioral strategies. Animals have developed a wide variety of adaptations to improve their chances of survival. Among the most successful strategies are camouflage, mimicry, speed, and strength, which help animals either avoid predators or become more effective hunters.

Camouflage: Blending with the Environment

Camouflage, or cryptic coloration, allows animals to blend in with their surroundings, making them harder to detect by both predators and prey. This survival strategy is used by both hunters and the hunted.

For instance, chameleons and cuttlefish are masters of camouflage, capable of changing their color and pattern to match their environment. These changes can happen in real-time, allowing them to remain undetected by predators or prey. Similarly, many insects, such as stick insects and leaf insects, have evolved to look almost identical to twigs or leaves, making it nearly impossible for predators to spot them.

Camouflage can also take the form of disruptive coloration, where an animal's patterns break up the outline of its body, confusing predators. The zebra's black-and-white stripes, for instance, may seem conspicuous at first glance, but in the tall grass and in large herds, the stripes create a visual confusion that makes it harder for predators like lions to single out one individual.

Mimicry: The Art of Deception

Mimicry is another evolutionary survival strategy, where one species evolves to resemble another. This can be used to avoid predation or even to enhance hunting success.

Batesian mimicry occurs when a harmless species evolves to imitate the appearance of a dangerous or unpalatable one. For example, the viceroy butterfly mimics the monarch butterfly, which is toxic to predators. As a result, predators avoid both species, even though the viceroy is harmless.

In contrast, aggressive mimicry is used by predators to deceive their prey. The anglerfish, for example, uses a bioluminescent lure that mimics the appearance of small prey animals. When unsuspecting prey approaches the lure, the anglerfish strikes with lightning speed.

Speed: Outrunning Danger

For many animals, speed is a crucial survival tool, particularly for prey species that rely on fleeing from predators. Gazelles, antelope, and hares are examples of animals that have evolved incredible speed and agility to escape from predators. Even large animals like zebras and wildebeest rely on speed and endurance to outrun lions and other large predators.

On the predator side, speed is equally important. Cheetahs, as mentioned earlier, are the fastest land animals, using their incredible acceleration to close the distance between them and their prey. Birds of prey, like the peregrine falcon, can dive at speeds of over 240 miles per hour, making them the fastest animals in the world when in pursuit of prey.

Strength: Power in the Wild

While some animals rely on speed or stealth, others use sheer strength to survive. Apex predators such as lions, tigers, and bears rely on their muscular power to overpower prey. Similarly, crocodiles and alligators use their incredibly powerful jaws to crush their victims.

Among herbivores, strength is often a defense mechanism. Elephants, rhinos, and buffaloes use their size and strength to ward off predators. Even smaller animals, such as wolverines and badgers, are renowned for their strength relative to their size, making them formidable opponents even for much larger predators.

Parental Care: Ensuring the Survival of Offspring

One of the most essential survival strategies in the animal kingdom is parental care, where adults invest time and energy in raising their young. While some species, like many fish and reptiles, produce hundreds or even thousands of offspring with little or no care, other species have developed complex parental behaviors to ensure their offspring's survival.

Nurturing the Young: A Vital Investment

In species that invest heavily in their young, parents may guard eggs, feed hatchlings, or protect offspring from predators. Mammals, in particular, are known for their extended periods of parental care. For example, a lioness will spend weeks hunting and bringing food to her cubs

Finding the Light

while teaching them the skills they need to survive as adults.

Birds are also excellent examples of parental care. Many bird species, like eagles and penguins, engage in cooperative parenting, where both parents take turns incubating eggs and feeding the young. The emperor penguin, for instance, endures the harsh Antarctic winter to incubate its egg on its feet, while the female penguin ventures to the ocean to feed.

Social Animals and Collective Care

In some species, the care of offspring is not limited to the parents. In social animals like wolves, elephants, and primates, the entire group or pack may help raise and protect the young. This strategy increases the chances of survival, as multiple adults are involved in feeding, teaching, and defending the offspring.

Elephants, for example, live in matriarchal herds where the older females help care for the calves. If a calf is in danger, the entire herd will rally to protect it. Similarly, in wolf packs, the entire group may take part in raising the pups, from feeding to teaching them hunting techniques.

Unusual Parenting Strategies

Finding the Light

Some species use more unusual methods to ensure their offspring's survival. The poison dart frog, for example, carries its tadpoles on its back, depositing them in small pools of water high in the trees. This reduces the chances of predation by fish or other aquatic creatures.

Another fascinating example is the mouthbrooding fish, which carries its young inside its mouth for protection. While this might seem dangerous, the fish ensures that its offspring are kept safe from predators until they are old enough to fend for themselves.

In the animal kingdom, the struggle for survival is an intricate dance of adaptation, cooperation, and competition. Predator and prey engage in a never-ending battle for supremacy, constantly evolving and improving their tactics for hunting or evading capture. Meanwhile, animals develop sophisticated strategies to ensure their own survival and that of their offspring, ensuring that life in the wild, though perilous, continues on its evolutionary path.

Chapter 5: The Human Spirit: Overcoming Adversity

The story of human resilience is as old as humanity itself. Throughout history, humans have faced wars, natural disasters, personal traumas, and countless other challenges that have threatened their survival. Yet, time and again, individuals and communities have shown an extraordinary ability to overcome adversity. What enables people to rise above the direst situations? Why do some survive and thrive, while others struggle to cope?

In this chapter, we will explore the concept of resilience — the mental, emotional, and physical strength that allows humans to endure hardship. We will examine how people have survived extreme situations, from shipwrecks to life-threatening illnesses, and how human innovation has played a crucial role in overcoming adversity. By looking at psychological, historical, and technological perspectives, we will gain a deeper understanding of the human capacity to persevere and grow through challenges.

The Psychology of Resilience

Resilience is the ability to bounce back from adversity, trauma, or hardship. While some people seem to naturally possess more resilience than others, research has shown that resilience is not a fixed trait. It is a skill that can be developed over time through mental and emotional practices.

Key Components of Resilience

1. **Optimism**: The belief that challenges can be overcome and that life will improve. Optimism doesn't mean ignoring problems, but rather approaching them with a solution-oriented mindset. It involves focusing on what can be controlled rather than what cannot.
2. **Self-Efficacy**: This refers to an individual's belief in their own ability to influence events and outcomes in their life. People who have high self-efficacy tend to take more proactive steps to solve problems, rather than feeling helpless in the face of adversity.
3. **Emotional Regulation**: Resilient people are better able to manage their emotions during stressful situations. They acknowledge their feelings but do not allow those emotions to overwhelm them or dictate their actions. This emotional balance helps in making rational decisions under pressure.
4. **Social Support**: Connection with others is a critical factor in resilience. Support from friends, family, and communities provides emotional comfort,

Finding the Light

practical help, and a sense of belonging. People who are part of a supportive network are more likely to recover from adversity.

5. **Sense of Purpose**: People who have a sense of purpose or meaning in life are often more resilient because they feel that they have something worth fighting for. Whether it's a personal mission, family, faith, or professional goals, a sense of purpose can help people endure difficult times.

Resilience in Action: Historical Examples

- Nelson Mandela: One of the most iconic examples of human resilience is Nelson Mandela, who endured 27 years in prison during South Africa's apartheid regime. Despite the harsh conditions, Mandela maintained hope and purpose, becoming a global symbol of perseverance. His mental strength, optimism, and sense of purpose not only allowed him to survive but to emerge as a leader who united a divided nation.
- **Viktor Frankl**: Viktor Frankl, a Holocaust survivor and psychiatrist, wrote about his experiences in Nazi concentration camps in his book *Man's Search for Meaning*. Frankl observed that those who survived were often the individuals who could find meaning in their suffering. His philosophy, called logotherapy, focuses on the idea that life has

Finding the Light

meaning under all circumstances, even the most miserable ones, and that finding that meaning helps people endure suffering.

Surviving Extreme Situations

Human history is filled with stories of survival against unimaginable odds. These tales not only highlight physical endurance but also the psychological and emotional strength that allows individuals to push beyond their limits. From shipwrecks to mountain expeditions, the human spirit has been tested time and again in extreme situations.

Shipwreck Survivors: The Story of the Essex

In 1820, the whaling ship *Essex* was struck by a sperm whale and sank in the Pacific Ocean. The crew of 20 men was left stranded in small boats, thousands of miles from land, with limited supplies. The ordeal, which lasted more than 90 days, pushed the men to the brink of despair. Some succumbed to hunger, thirst, and exposure, while others resorted to cannibalism to survive.

The survivors of the *Essex* demonstrated extraordinary resilience. They rationed food and water, navigated the vast ocean using limited knowledge, and remained determined to stay alive despite overwhelming odds. Their story, which later inspired Herman Melville's *Moby-Dick*, is a testament to the human will to survive, even when faced with the harshest conditions imaginable.

The Andes Flight Disaster: The Miracle of the Andes

In 1972, a Uruguayan rugby team's plane crashed into the Andes Mountains, leaving 16 survivors stranded in one of the most inhospitable environments on Earth. For more than two months, the group battled freezing temperatures, avalanches, and starvation. With no hope of rescue, they were forced to make the agonizing decision to eat the bodies of their deceased friends to stay alive.

Despite the unimaginable horrors they faced, the survivors maintained hope and worked together to stay alive. Two of the survivors eventually trekked for ten days through the mountains to find help, leading to the rescue of the remaining survivors. Their story became the basis for the book *Alive*, which explores the psychological and physical resilience required to survive such a harrowing ordeal.

Mount Everest: The 1996 Disaster

Mount Everest, the highest peak on Earth, has long been a symbol of human ambition and resilience. In 1996, a deadly blizzard struck the mountain, leading to the deaths of eight climbers. However, many others survived, despite being exposed to the elements for hours in what should have been fatal conditions.

One of the most remarkable survival stories from that expedition is that of Beck Weathers, a climber who was left for dead twice during the blizzard. Weathers suffered severe frostbite and lost several fingers, toes, and part of his face. Yet, he managed to stagger back to camp, driven by sheer willpower and the desire to see his family again.

Innovation in Adversity

While the human spirit is an essential component of resilience, innovation and adaptability have also played a crucial role in helping individuals and societies overcome adversity. Throughout history, humans have developed new technologies, medicines, and community practices that have allowed them to survive and thrive in the face of challenges.

Technology and Survival

- **The Wheel and Transportation**: One of humanity's earliest innovations, the wheel, transformed how humans moved goods and traveled across distances. The ability to transport resources more efficiently

helped early civilizations survive harsh environments, droughts, and long migrations.
- **Agricultural Innovation**: Agriculture has been another key to human survival. Innovations such as crop rotation, irrigation, and the domestication of animals allowed humans to settle in one place, leading to the rise of civilizations. These practices helped societies weather famines and environmental changes.
- **Modern Medicine**: Advances in medicine have played a significant role in increasing human resilience. From antibiotics to vaccines, medical innovations have allowed people to survive diseases and injuries that would have been fatal in the past. The eradication of smallpox, the development of life-saving surgeries, and the creation of emergency care systems have dramatically improved human survival rates.

Social and Community Adaptations

- Community Resilience: In many cases, resilience is not just an individual trait but a communal one. Communities often come together to face adversity, whether it's rebuilding after natural disasters or supporting each other through economic hardships. Social networks provide emotional support,

resources, and collective action, all of which contribute to resilience.
- **Cultural Innovations**: Human cultures have developed unique ways of dealing with adversity. For example, in Japan, the concept of *gaman*, which means enduring hardship with patience and dignity, has helped individuals and communities navigate disasters such as earthquakes and tsunamis. In other cultures, religious and spiritual practices provide solace and a framework for understanding suffering, which in turn fosters resilience.
- **Post-War Rebuilding**: After World War II, countries devastated by the conflict, such as Germany and Japan, rebuilt their economies and infrastructure through innovative strategies. The Marshall Plan, initiated by the United States, provided financial aid and technological expertise to help war-torn Europe recover. These efforts not only restored nations but also fostered resilience and growth in the aftermath of destruction.

Conclusion: The Human Capacity to Overcome

The human spirit's ability to endure and overcome adversity is a remarkable trait. Whether through psychological resilience, physical endurance, or innovation, humans have consistently found ways to survive and thrive in the face of extreme challenges. The stories of survival, from shipwrecks to mountain disasters, highlight the power of hope, determination, and the will to live.

In the modern world, where we continue to face global challenges like pandemics, climate change, and political instability, the lessons of resilience remain as relevant as ever. By understanding the mental, emotional, and social tools that enable us to persevere, we can continue to find the light even in the darkest moments of life.

Chapter 6: Triumph of the Will: Stories of Survival

Throughout history, individuals have faced unimaginable challenges that test the boundaries of human endurance and resilience. These stories of survival serve as powerful reminders of life's capacity to triumph over adversity. Whether in the face of natural disasters, war, or personal illness, the human spirit has demonstrated an extraordinary will to endure and prevail. This chapter delves into the remarkable stories of those who have survived through sheer determination, resourcefulness, and inner strength. Their experiences not only inspire but also shed light on the incredible resilience that defines life.

Survivors of Natural Disasters

Natural disasters, with their unpredictable and often cataclysmic force, have tested humanity for centuries. Earthquakes, tsunamis, hurricanes, and volcanic eruptions can decimate entire communities, leaving survivors to grapple with the physical, emotional, and psychological aftermath. Yet, amid the devastation, there are remarkable stories of individuals who have defied the odds and found ways to survive the most harrowing situations.

The Chilean Miners: Buried Alive for 69 Days

Finding the Light

In 2010, one of the most miraculous survival stories emerged from the Atacama Desert in Chile. A group of 33 miners became trapped 2,300 feet underground after a cave-in at the San José mine. With no immediate means of rescue and limited supplies, the miners faced the terrifying prospect of slow suffocation or starvation in the dark, confined space of the mine. However, their survival was not merely a matter of luck; it was a testament to human resilience and unity.

The miners, led by their foreman Luis Urzúa, organized themselves into a disciplined and cooperative group. They rationed food and water meticulously, creating a makeshift survival plan that would extend their meager supplies. In the face of despair, they maintained hope and camaraderie, bolstering one another's spirits. It took over two weeks before rescuers managed to drill a hole large enough to send food, water, and communication devices to the men. Finally, after 69 days underground, all 33 miners were pulled to safety, their will to survive having carried them through the ordeal.

Tsunami Survivor Petra Nemcova: Finding Light After Tragedy

The 2004 Indian Ocean tsunami, one of the deadliest natural disasters in history, claimed the lives of over 230,000 people across 14 countries. Among those affected

Finding the Light

was Czech supermodel Petra Nemcova, who was vacationing in Thailand with her fiancé, Simon Atlee. When the tsunami struck, it swept away buildings, people, and entire communities in an instant. Nemcova was seriously injured, breaking her pelvis in multiple places, but she managed to survive by clinging to a palm tree for eight hours as waves and debris crashed around her. Tragically, her fiancé did not survive.

Nemcova's survival story is not only one of physical endurance but also emotional resilience. In the years following the disaster, she channeled her grief and trauma into creating the Happy Hearts Fund, an organization dedicated to rebuilding schools in areas affected by natural disasters. Her work has helped thousands of children regain access to education and hope, demonstrating how survival is not just about enduring trauma, but also about finding meaning and purpose in its aftermath.

Survival in War

War has always been one of the most devastating forms of human conflict. The physical and psychological toll it takes on soldiers and civilians alike is often unimaginable. Yet, in the midst of war's horrors, there are countless stories of individuals who have survived through sheer willpower and ingenuity, refusing to surrender to the chaos around them.

Finding the Light

Louis Zamperini: An Unbroken Spirit

Louis Zamperini's story is one of the most extraordinary examples of survival in war. An Olympic athlete turned U.S. airman during World War II, Zamperini's life took a harrowing turn when his bomber crashed into the Pacific Ocean during a mission. He and two other crew members spent 47 days adrift in a lifeboat, enduring extreme hunger, thirst, and the threat of shark attacks. One of the men died, but Zamperini and his fellow survivor, Phil Phillips, managed to stay alive by catching rainwater and fishing.

Zamperini's ordeal didn't end there. After being captured by Japanese forces, he spent over two years in brutal prisoner-of-war camps, where he was tortured and beaten mercilessly, particularly by a sadistic guard known as "The Bird." Despite the suffering, Zamperini refused to give up. He clung to the belief that he would one day return home, and his mental resilience helped him endure the unimaginable. After the war, Zamperini's story of survival was immortalized in Laura Hillenbrand's best-selling book *Unbroken*, and he went on to become a symbol of the unbreakable human spirit.

The Siege of Leningrad: Survival Against All Odds

The Siege of Leningrad, which lasted from 1941 to 1944 during World War II, is one of the longest and most brutal sieges in history. For nearly 900 days, the citizens of

Leningrad (modern-day Saint Petersburg) endured relentless bombardment and starvation as the Nazi forces attempted to starve the city into submission. With supply lines cut off, food became scarce, and the residents were forced to survive on meager rations, often resorting to eating rats, pets, and even wallpaper paste.

Despite these horrific conditions, the people of Leningrad displayed an incredible will to survive. They organized communal kitchens, set up makeshift hospitals, and continued cultural activities, including music and art, to maintain their morale. One of the most famous symbols of the city's resilience was the performance of Dmitri Shostakovich's *Leningrad Symphony* in August 1942, which took place in the besieged city despite the ongoing bombardment. The siege eventually ended in Soviet victory, and the survival of the city's inhabitants remains a powerful testament to human endurance in the face of overwhelming odds.

Personal Triumphs

While natural disasters and wars are extreme tests of survival, personal adversity can also push individuals to their limits. Illness, disability, and life-threatening conditions often demand the same level of courage, determination, and resilience. The following stories illustrate how, even when faced with the most daunting challenges, the human spirit can find a way to overcome.

Helen Keller: Defying the Darkness and Silence

Helen Keller's story is one of the most remarkable examples of personal triumph over adversity. Born in 1880, Keller lost both her sight and hearing due to an illness when she was just 19 months old. For the first few years of her life, she was trapped in a world of darkness and silence, unable to communicate or understand the world around her. However, with the help of her teacher, Anne Sullivan, Keller learned to communicate through tactile sign language, opening up an entirely new world of possibilities.

Keller went on to become the first deaf-blind person to earn a college degree, graduating from Radcliffe College in 1904. She became an author, activist, and advocate for people with disabilities, as well as a champion for women's suffrage and workers' rights. Keller's life is a testament to the power of perseverance and the human capacity to transcend even the most formidable barriers.

Aron Ralston: Surviving the Impossible

In 2003, Aron Ralston, an experienced outdoorsman, faced a life-or-death situation while hiking alone in Utah's Blue John Canyon. A boulder became dislodged and pinned

Ralston's arm against the canyon wall, trapping him in a remote area with little hope of rescue. After five days of being trapped, with his water supply exhausted and realizing that no one was coming to save him, Ralston made the unimaginable decision to amputate his own arm using a dull multi-tool.

Ralston's extraordinary act of survival became the subject of the film *127 Hours*. His story is a powerful example of how far humans will go to preserve life and the indomitable will to survive, even in the most desperate of circumstances.

Stephen Hawking: A Mind Unbound by Disability

Physicist Stephen Hawking is another iconic figure whose life exemplifies personal triumph over extreme adversity. Diagnosed with amyotrophic lateral sclerosis (ALS), or Lou Gehrig's disease, at the age of 21, Hawking was given just a few years to live. Despite the grim prognosis, he defied the odds and lived for more than five decades with the debilitating illness, which gradually paralyzed him.

Unable to move or speak without assistance, Hawking continued to revolutionize the field of theoretical physics, making groundbreaking contributions to our understanding of black holes, cosmology, and the nature of the universe. His book, *A Brief History of Time*, became a global bestseller, and he inspired millions as a symbol of the limitless potential of the human mind.

These stories of survival, whether in the face of natural disasters, war, or personal challenges, illustrate the profound resilience that defines life. The ability to endure, adapt, and ultimately triumph over adversity speaks to the incredible strength that lies within every individual. These examples remind us that no matter how dark the situation, there is always light to be found, and life will always find a way.

Chapter 7: Nature's Recovery After Disasters

Nature's resilience is one of its most awe-inspiring qualities. From wildfires that reduce entire forests to ash, to volcanic eruptions that bury landscapes in lava and ash, to the bleaching of coral reefs that seemingly transforms vibrant underwater ecosystems into ghostly wastelands, the natural world has an incredible capacity to heal and regenerate. Life, it seems, has a built-in mechanism for recovery that ensures its continuity, even in the face of catastrophic events. In this chapter, we explore how nature finds its way back to health after disasters, highlighting the remarkable processes of **forest regeneration after wildfires**, the **renewal of life after volcanic eruptions**, and the **recovery of coral reefs after bleaching events**.

Forest Regeneration: How Ecosystems Bounce Back After Wildfires

Wildfires, while destructive and often feared, play an essential role in the life cycle of many ecosystems, particularly in forests. Fires can consume everything in their path, leaving behind a charred and seemingly lifeless landscape. However, many forests are adapted to fire and rely on it as part of their natural regeneration process.

The Role of Fire in Forest Ecosystems

In fire-prone environments like the coniferous forests of the western United States, the Mediterranean woodlands, and savannas, fire is a natural part of the ecosystem's rhythm. For centuries, these ecosystems have evolved to not only withstand fire but to use it as a tool for renewal. Some species of trees and plants, such as the **lodgepole pine** and **sequoias**, have even developed mechanisms to require fire for regeneration. The heat from a wildfire triggers the opening of their resin-coated cones, releasing seeds that fall onto the nutrient-rich ash bed below, providing a fertile ground for new growth.

Ecological Benefits of Wildfires:

1. **Clearing Underbrush**: Fires remove dense undergrowth, dead trees, and other debris, which can otherwise fuel larger, more devastating fires. This clearing also allows more sunlight to reach the forest floor, creating a better environment for new plants to grow.
2. **Nutrient Cycling**: The burning of organic matter releases nutrients back into the soil, making it more fertile and conducive to plant life. Phosphorus, potassium, and other minerals that were locked in dead organic matter become accessible, stimulating new plant growth.

3. **Stimulating Growth**: For certain plant species, fire is a crucial trigger for germination. Some seeds are encased in hard shells that require the heat of a fire to crack open, allowing them to sprout. Others may have fire-activated chemical signals that induce growth.

The Recovery Process

Once a fire has passed, the initial scene is one of destruction. Trees may be blackened stumps, and the ground may appear scorched and lifeless. Yet, just beneath the surface, life is already preparing to return.

1. **Pioneer Species**: The first plants to recolonize a burned area are known as pioneer species. These are typically fast-growing grasses, herbs, and shrubs that can quickly take advantage of the newly cleared space. Species like **fireweed** and **lupines** are common examples of plants that thrive after fires. They stabilize the soil, prevent erosion, and pave the way for the return of larger plants and trees.
2. **Resilient Trees**: Fire-adapted tree species, such as certain pines, sprout new growth from their bases or germinate from seeds released during the fire. These new saplings are better suited to the post-fire environment, as they grow in the nutrient-rich soil left behind by the blaze.

3. **Wildlife Return**: As vegetation returns, so do the animals. Herbivores like deer and rabbits are often the first to repopulate recovering forests, attracted by the fresh growth of grasses and shrubs. Predators soon follow, and the ecosystem slowly rebalances itself. Birds such as woodpeckers are often among the first to return, feeding on the insects that thrive in the dead wood of fire-damaged trees.

Over time, the forest rebuilds itself, often even more diverse and resilient than before. In some cases, a burned forest will transition into a new ecosystem, depending on the severity of the fire and the climate conditions that follow.

Case Study: Yellowstone National Park Wildfires of 1988

One of the most well-known examples of forest regeneration occurred after the Yellowstone National Park fires of 1988. These fires, which consumed nearly 800,000 acres of the park, were initially seen as a devastating ecological disaster. However, within just a few years, signs of life began to reappear. Lodgepole pines, which had evolved to rely on fire for regeneration, began sprouting en masse, and the park's biodiversity quickly rebounded. The Yellowstone fires are now a classic example of how nature can turn destruction into renewal.

Rebirth After Eruptions: The Renewal of Life Around Volcanoes Like Mount St. Helens

Volcanic eruptions are among the most powerful and destructive forces in nature. They can obliterate landscapes, burying them under layers of lava and ash, and altering ecosystems in an instant. However, volcanic eruptions also create opportunities for life to flourish in new and unique ways. The rich volcanic soil left behind after an eruption is full of minerals that are highly beneficial to plant life, and in time, these barren landscapes can become hotspots of biodiversity.

The Aftermath of Volcanic Eruptions

When a volcano erupts, the immediate area is often rendered lifeless. Lava flows incinerate everything in their path, and thick layers of ash can smother plants and animals. The blast itself can knock down entire forests and create a wasteland of debris. Yet, even in this seemingly apocalyptic environment, life finds a way to return.

Mount St. Helens: A Case Study in Regeneration

The eruption of **Mount St. Helens** in 1980 is one of the most iconic examples of nature's recovery after a volcanic disaster. The eruption was massive, leveling over 230

Finding the Light

square miles of forest, killing thousands of animals, and leaving a thick layer of volcanic ash across the landscape. The destruction seemed total. Yet, within a few years, life began to reassert itself.

1. **Primary Succession**: After the eruption, the area around Mount St. Helens became a laboratory for scientists studying **primary succession**, the process by which life returns to a completely sterile environment. Initially, it seemed that the ash-covered landscape would remain barren for decades, but within months, hardy pioneer species such as **lupines** and mosses began to colonize the area. These plants helped to break down the volcanic material, creating soil that could support more complex life.
2. **Animal Return**: As plants began to take root, animals followed. Small mammals like **pocket gophers** were among the first to return, digging through the ash and bringing organic matter to the surface, which further enriched the soil. Birds, particularly seed-eaters like sparrows and finches, soon arrived, feeding on the plants that had begun to grow.
3. **Lakes and Waterways**: The eruption also created new lakes and altered existing water systems. Over time, these bodies of water became home to fish,

amphibians, and aquatic plants, helping to further restore the ecosystem.

The Long-Term Effects

Today, more than four decades after the eruption, the area around Mount St. Helens is a thriving ecosystem. The once-devastated landscape is now covered in lush vegetation, and wildlife has returned in abundance. The eruption, while catastrophic, created new opportunities for life, leading to a more diverse and dynamic ecosystem than existed before.

Coral Reefs After Bleaching: Efforts to Restore and Preserve These Vital Ecosystems

Coral reefs are among the most diverse and vital ecosystems on Earth, home to a quarter of all marine life despite covering less than 1% of the ocean floor. However, these vibrant ecosystems are incredibly sensitive to changes in water temperature, pollution, and ocean acidification. When corals are stressed by rising temperatures or environmental changes, they expel the symbiotic algae living in their tissues, causing them to turn white—a phenomenon known as **coral bleaching**. While bleached coral is not immediately dead, it is weakened and more vulnerable to disease and death.

Finding the Light

The Impact of Coral Bleaching

Coral bleaching events have become more frequent and severe due to climate change. When water temperatures rise even slightly, corals lose their algae, which provide them with energy through photosynthesis. Without this energy, corals can starve and die. In recent decades, large-scale bleaching events have devastated coral reefs around the world, including the **Great Barrier Reef**.

Efforts to Restore Coral Reefs

Despite the dire outlook for many coral reefs, efforts are underway to help reefs recover and adapt to changing conditions.

1. **Coral Nurseries**: One of the most promising methods of reef restoration is the establishment of **coral nurseries**, where healthy coral fragments are grown in controlled environments and then transplanted onto damaged reefs. These nurseries allow scientists to grow resilient species that are more tolerant of warmer waters. Once these corals are reintroduced to the wild, they can help repopulate and restore damaged ecosystems.
2. **Selective Breeding**: Scientists are also experimenting with **selective breeding** of corals, choosing individuals that have shown resilience to bleaching events and breeding them to create more

Finding the Light

resistant strains. These "super corals" are then introduced to vulnerable areas to strengthen the overall health of the reef.

3. **Reducing Local Stressors**: While global warming is a major cause of coral bleaching, local factors such as overfishing, pollution, and coastal development also contribute to reef decline. Conservation efforts aimed at reducing these local stressors—such as creating marine protected areas, regulating fishing practices, and improving water quality—can give reefs a better chance of recovery.
4. **Assisted Evolution**: Another cutting-edge approach is **assisted evolution**, where scientists accelerate the natural evolutionary processes of coral. By exposing corals to stressors in a controlled environment, researchers can speed up their adaptation to warmer, more acidic oceans.

Hope for the Future

While coral reefs face unprecedented threats, the resilience of nature should not be underestimated. With concerted conservation efforts, it is possible to restore and protect these vital ecosystems. Initiatives like those in the Great Barrier Reef, where coral transplantation and selective breeding are showing promising results, offer hope that these underwater rainforests can continue to thrive for generations to come.

Conclusion

Nature's ability to recover from disasters is nothing short of miraculous. Whether through the regeneration of forests after wildfires, the rebirth of life after volcanic eruptions, or the ongoing efforts to restore coral reefs after bleaching events, the natural world demonstrates time and time again that it can overcome even the most severe challenges. While human intervention can play a role in aiding this recovery, it is nature's inherent resilience that remains at the heart of every comeback story. Understanding these processes not only deepens our appreciation for the natural world but also reinforces the importance of protecting and preserving it for the future.

Chapter 8: Microbial Resilience – The Hidden World of Survival

Microorganisms, such as bacteria, viruses, fungi, and archaea, are among the most resilient forms of life on Earth. Although invisible to the naked eye, they play crucial roles in maintaining ecosystems, human health, and even the evolutionary processes of life. These organisms exhibit astonishing survival capabilities, thriving in some of the most extreme conditions on the planet. Their ability to adapt rapidly to changing environments ensures their continued existence, even in the face of human interventions like antibiotics and antivirals. In this chapter, we delve into the hidden world of microbial resilience, exploring how these microscopic organisms shape life on Earth and defy the odds of survival.

Microbial Extremophiles: Bacteria Thriving in Boiling Springs and Deep-Sea Trenches

Extremophiles are microorganisms that flourish in conditions once thought to be uninhabitable. They survive in environments that are too hot, too cold, too acidic, too alkaline, or too salty for most other life forms. These organisms, particularly certain bacteria and archaea, offer a

Finding the Light

glimpse into the tenacity of life and its ability to adapt to the most extreme conditions imaginable.

Thermophiles: Life in Boiling Springs

Thermophiles are a class of extremophiles that thrive in extremely high temperatures. These heat-loving organisms are found in environments like hydrothermal vents and geothermal springs, where temperatures can reach upwards of 100°C (212°F). One of the most famous examples of thermophiles is *Thermus aquaticus*, a bacterium first discovered in the hot springs of Yellowstone National Park. This organism can survive and reproduce in temperatures that would be lethal to most life forms.

Thermus aquaticus became world-renowned for its role in the polymerase chain reaction (PCR), a revolutionary technique in molecular biology. The enzyme Taq polymerase, derived from *Thermus aquaticus*, can withstand the high temperatures necessary for DNA replication in PCR, enabling rapid advances in genetics and medicine. The discovery of this heat-resistant enzyme highlights the incredible adaptability of thermophiles and their profound impact on human scientific progress.

Barophiles: Survival in Deep-Sea Trenches

At the opposite end of the temperature spectrum, barophiles (also known as piezophiles) are organisms that

Finding the Light

thrive under extreme pressure, such as those found in deep-sea trenches. These microorganisms live at depths of over 11,000 meters (36,000 feet), where pressures can exceed 1,000 times that of the atmosphere at sea level.

Barophilic bacteria have adapted to these crushing pressures by altering the composition of their cell membranes and proteins. These changes allow their cellular machinery to function despite the immense forces exerted by the surrounding water. The Mariana Trench, the deepest part of the world's oceans, harbors diverse microbial communities that have evolved to survive in this dark, cold, and high-pressure environment. These bacteria play essential roles in nutrient cycling and decomposition in the deep ocean, driving the biogeochemical processes that sustain life even in the planet's most remote ecosystems.

The existence of extremophiles like thermophiles and barophiles not only expands our understanding of the limits of life on Earth but also raises the possibility of life beyond our planet. The discovery of microbial life in extreme environments on Earth has led astrobiologists to speculate that similar organisms could exist on other celestial bodies, such as Mars or the icy moons of Jupiter and Saturn.

Antibiotic Resistance: How Bacteria Evolve Rapidly to Resist Human Efforts to Control Them

While extremophiles showcase the adaptability of microorganisms to physical extremes, antibiotic-resistant bacteria demonstrate their ability to evolve in response to human intervention. The discovery of antibiotics in the early 20th century revolutionized medicine, offering effective treatments for bacterial infections that once caused millions of deaths. However, the overuse and misuse of antibiotics have led to the rise of antibiotic-resistant bacteria, which pose a growing threat to global health.

Mechanisms of Resistance

Bacteria are remarkable for their ability to evolve rapidly, especially under selective pressure from antibiotics. Resistance mechanisms can develop through various means, including genetic mutations and the acquisition of resistance genes from other bacteria via horizontal gene transfer. These mechanisms allow bacteria to neutralize antibiotics, making them ineffective.

Some of the most common resistance strategies include:

- **Efflux Pumps:** Certain bacteria develop proteins that actively pump antibiotics out of their cells before the drugs can exert their effects.

Finding the Light

- **Enzyme Production:** Bacteria can produce enzymes that break down antibiotics. For instance, beta-lactamases are enzymes that degrade beta-lactam antibiotics, such as penicillin and cephalosporins, rendering them useless.
- **Altered Target Sites:** Some bacteria mutate the specific targets of antibiotics within their cells, preventing the drugs from binding to and disabling these critical sites.
- **Biofilm Formation:** Bacterial communities can form biofilms—thick layers of cells encased in a protective matrix—that shield them from antibiotic exposure and immune system attacks.

The Rise of Superbugs

The overuse of antibiotics in healthcare, agriculture, and animal husbandry has accelerated the emergence of so-called "superbugs"—bacteria that are resistant to multiple classes of antibiotics. Some well-known superbugs include:

- **Methicillin-Resistant Staphylococcus aureus (MRSA):** MRSA is a strain of *Staphylococcus aureus* that has developed resistance to methicillin and other commonly used antibiotics. It is a significant cause of hospital-acquired infections.
- **Carbapenem-Resistant Enterobacteriaceae (CRE):** CRE are a group of bacteria that have

developed resistance to carbapenems, which are considered antibiotics of last resort.
- **Multi-Drug Resistant Tuberculosis (MDR-TB):** MDR-TB is a form of tuberculosis caused by bacteria resistant to at least two of the most potent TB drugs.

The rise of antibiotic-resistant bacteria threatens to undermine the effectiveness of modern medicine. Without effective antibiotics, routine surgeries, cancer treatments, and organ transplants could become life-threatening due to the risk of untreatable infections. To combat this growing crisis, researchers are exploring alternative therapies, such as phage therapy (using viruses that infect and kill bacteria) and the development of new classes of antibiotics.

Viruses in the Natural World: Their Role in Evolution and Ecological Balance

While bacteria and archaea demonstrate impressive resilience in adapting to physical and chemical extremes, viruses—often seen as agents of disease—play an equally critical role in the natural world. Viruses, which are not technically "alive" because they require a host to replicate, are nonetheless a driving force in evolution and ecological balance.

Viral Impact on Evolution

Viruses are constantly interacting with their hosts, driving evolutionary changes through processes such as genetic recombination and horizontal gene transfer. One of the most significant evolutionary impacts of viruses is their role in transferring genes between different organisms. Known as **horizontal gene transfer**, this process allows for the rapid acquisition of new traits, contributing to the evolution of species over time.

For example, retroviruses, a class of viruses that includes HIV, insert their genetic material into the host's genome during infection. Over millions of years, remnants of ancient viral infections have been incorporated into the DNA of various organisms, including humans. In fact, scientists estimate that up to 8% of the human genome consists of viral DNA, known as **endogenous retroviruses**. Some of these viral genes have been repurposed by the host organism for beneficial functions, including the development of the placenta in mammals.

Viruses as Ecological Regulators

In addition to their role in evolution, viruses are important regulators of ecosystems, particularly in aquatic environments. Marine viruses, for instance, infect and kill bacteria and phytoplankton in the ocean, influencing nutrient cycling and the global carbon cycle. By lysing (breaking open) bacterial cells, marine viruses release organic matter back into the environment, providing nutrients for other organisms in the food web. This process, known as the **viral shunt**, prevents carbon from sinking to the deep ocean, keeping it available for use in surface waters.

Viruses also play a role in controlling population dynamics within ecosystems. For example, outbreaks of certain viral infections can limit the population size of a particular species, preventing overpopulation and maintaining ecological balance. This regulatory function ensures that no single species dominates an ecosystem, allowing for biodiversity and stability.

Conclusion

Microbial resilience is a testament to life's adaptability and tenacity, even in the face of extreme environmental conditions and human intervention. From extremophiles that thrive in boiling springs and deep-sea trenches to antibiotic-resistant bacteria that challenge modern medicine, microorganisms have evolved sophisticated survival strategies that make them some of the most formidable life forms on Earth. Viruses, despite their association with disease, also play crucial roles in driving evolution and regulating ecosystems, showcasing their

Finding the Light

importance in the broader web of life. In the hidden world of microbes, resilience is not just about survival—it is about thriving and shaping the very fabric of life on our planet.

Chapter 9: Healing and Rebirth in Ecosystems

Ecosystems are dynamic, interconnected webs of life that continuously experience cycles of growth, disturbance, destruction, and renewal. The delicate balance of an ecosystem can be disrupted by both natural phenomena, such as wildfires and floods, and human activities like deforestation, pollution, and urban development. Yet despite the magnitude of these disruptions, ecosystems have a remarkable capacity for healing and rebirth. This resilience is a testament to the adaptability of life on Earth and the intricate relationships within ecosystems that foster recovery and regrowth.

This chapter explores the processes that enable ecosystems to rebound after disturbances, the critical role of keystone species in maintaining ecological balance, and how human intervention can support the healing of ecosystems that have been pushed to the brink of collapse.

Ecosystem Resilience: How Ecosystems Recover from Disturbances

Ecosystem resilience refers to the ability of an ecosystem to withstand, absorb, and recover from disturbances while retaining its essential functions and structure. Different ecosystems, from forests to wetlands, oceans to grasslands, each exhibit unique mechanisms of resilience that allow them to respond to environmental changes.

Natural Disturbances and Ecosystem Recovery

Natural disturbances, such as wildfires, hurricanes, floods, and volcanic eruptions, are part of the Earth's ecological rhythm. While these events may seem catastrophic, they often play essential roles in maintaining the health and diversity of ecosystems. For example:

- **Wildfires in Forest Ecosystems:** In many forest ecosystems, wildfires are not just destructive forces but agents of renewal. Fires can clear out dead wood, underbrush, and diseased plants, making space for new growth. Certain species, such as the lodgepole pine, rely on fire for reproduction, as the heat from the fire opens their cones, releasing seeds into the nutrient-rich ash. Over time, forests that are adapted to regular fire cycles can bounce back even stronger, with greater biodiversity.
- **Floods in Wetlands and River Systems:** Flooding, though often destructive in human terms, is crucial for wetlands and floodplain ecosystems. Floodwaters

spread nutrient-rich silt across the land, fertilizing the soil and promoting plant growth. They also replenish water tables and create new habitats for aquatic species. Wetland ecosystems are particularly resilient to flooding because many of the plant species are adapted to waterlogged conditions and can thrive in the aftermath of flood events.

- **Volcanic Eruptions and Island Formation:** Volcanic eruptions can obliterate landscapes, yet they also create new ones. Islands formed by volcanic activity, like those in the Galápagos, provide blank slates for ecological succession, where species slowly colonize the barren land. Over time, plants, insects, and birds establish new ecosystems, demonstrating nature's ability to build life from the ground up.

Human-Induced Disturbances

While ecosystems are equipped to handle natural disturbances, human activities often cause more severe and long-lasting damage. Deforestation, pollution, oil spills, and overfishing can push ecosystems beyond their natural ability to recover. However, even in the face of such disruptions, ecosystems can regenerate if given the opportunity.

- **Deforestation and Forest Recovery:** Deforestation, driven by agriculture, logging, and urban development, is one of the most significant threats to forests worldwide. Yet, when left undisturbed, secondary forests can regenerate over time. This process of natural reforestation can take decades or even centuries, depending on the severity of the damage and the ecosystem's capacity for regrowth. In some cases, replanting efforts and habitat restoration projects can accelerate the recovery process.
- **Oil Spills in Marine Ecosystems:** Oil spills are particularly devastating to marine ecosystems, smothering coral reefs, poisoning marine life, and destroying habitats. Despite this, ecosystems have a remarkable ability to recover, given time and intervention. For example, after the 2010 Deepwater Horizon oil spill in the Gulf of Mexico, concerted efforts were made to clean up the oil, restore damaged habitats, and protect vulnerable species. Marine life, including fish populations and coral reefs, began to show signs of recovery within years, though full restoration may take decades.

The Role of Keystone Species: Guardians of Ecological Balance

Finding the Light

A key factor in ecosystem resilience is the presence of **keystone species**. These species play a disproportionately large role in maintaining the structure and function of their ecosystems. Their presence or absence can have a cascading effect on the entire ecosystem, influencing everything from species diversity to nutrient cycling. Keystone species act as stabilizers, helping ecosystems recover after disturbances by ensuring that critical processes continue to function.

What Are Keystone Species?

Keystone species are organisms whose impact on their environment is much greater than their abundance or biomass would suggest. They can be predators, herbivores, or even plants, but their common feature is their role in maintaining ecological balance. When a keystone species is removed, the entire ecosystem can be thrown into disarray, leading to a loss of biodiversity and the collapse of the ecological network.

Examples of Keystone Species

- **Wolves in Yellowstone National Park:** One of the most famous examples of a keystone species is the reintroduction of wolves to Yellowstone National Park in the 1990s. Before the wolves' return, the park's ecosystem had become unbalanced, with elk populations growing out of control and overgrazing

Finding the Light

the vegetation. This overgrazing led to soil erosion, the destruction of riverbanks, and the loss of habitats for other species. When wolves were reintroduced, they helped control the elk population, allowing vegetation to recover. As a result, riverbanks stabilized, beavers returned to build dams, and biodiversity increased across the park.

- **Sea Otters in Kelp Forests:** Sea otters play a crucial role in maintaining the health of kelp forests along the Pacific coast. By feeding on sea urchins, which can decimate kelp if left unchecked, sea otters prevent the overgrazing of these underwater forests. In areas where sea otters have been removed, kelp forests have collapsed, leading to a loss of biodiversity and habitat for marine life. When sea otters are reintroduced, the ecosystem rebounds, with kelp forests regrowing and marine species returning.
- **Elephants in African Savannas:** Elephants are keystone species in African savannas due to their role in shaping the landscape. As they travel and feed, elephants knock down trees and clear vegetation, which maintains the balance between grasslands and forests. This, in turn, supports a diverse range of species, from grazing herbivores to predators. Without elephants, savannas would

become overgrown, reducing habitat diversity and leading to the decline of many species.

Human Intervention: How Conservation Efforts Are Helping Endangered Ecosystems Find New Light

While ecosystems have a natural ability to heal, human intervention is often necessary to aid in recovery, especially when the damage is severe or the natural processes of renewal are hindered. Conservation efforts around the world have demonstrated that, with the right strategies, it is possible to restore ecosystems to health and protect endangered species.

Habitat Restoration

Habitat restoration is a critical component of conservation efforts. By replanting native vegetation, removing invasive species, and restoring natural water flows, conservationists can help ecosystems regain their balance. Some examples include:

- **Wetland Restoration Projects:** Wetlands are among the most productive ecosystems on Earth, providing habitat for countless species and playing a vital role in water purification and flood control. However, many wetlands have been drained for agriculture or urban development. Restoration projects that re-flood wetlands and replant native vegetation have led to the return of wildlife and the re-establishment of ecosystem functions.

- **Coral Reef Restoration:** Coral reefs are particularly vulnerable to climate change, pollution, and overfishing. Coral restoration projects involve growing coral fragments in nurseries and transplanting them back onto damaged reefs. These efforts, combined with the reduction of local stressors like pollution and destructive fishing practices, have shown promise in helping coral reefs recover.

Legal Protections and Wildlife Conservation

In many cases, legal protections are necessary to prevent further damage to ecosystems. National parks, marine reserves, and wildlife sanctuaries provide safe havens for endangered species and allow ecosystems to recover without the pressure of human exploitation.

- **Marine Protected Areas (MPAs):** MPAs are zones where fishing, mining, and other extractive activities are restricted or prohibited. These areas give marine ecosystems a chance to regenerate, allowing fish populations to rebound and habitats like coral reefs to recover. Studies have shown that MPAs not only

Finding the Light

benefit the protected areas but also help surrounding regions by increasing fish stocks and biodiversity.
- **Endangered Species Protection:** Many species that play crucial roles in their ecosystems are at risk of extinction. Conservation efforts aimed at protecting endangered species, such as breeding programs, habitat preservation, and anti-poaching initiatives, are essential for the survival of these keystone species and the ecosystems they support.

Community-Led Conservation

Increasingly, conservation efforts are recognizing the importance of involving local communities in ecosystem restoration. Indigenous knowledge, sustainable resource management practices, and community engagement can all contribute to successful conservation outcomes.

- **Community-Based Forest Management:** In countries like Nepal and Brazil, community-based forest management programs have empowered local people to take charge of forest restoration and conservation. By giving communities a stake in the health of their ecosystems, these programs have led to the recovery of degraded forests and the sustainable use of resources.
- **Ecotourism and Conservation:** Ecotourism, when managed responsibly, can provide economic

incentives for conservation. In regions like Costa Rica and Tanzania, ecotourism has helped fund conservation projects while providing income to local communities. By creating economic value around biodiversity, ecotourism helps ensure that ecosystems are protected and preserved for future generations.

Conclusion

Ecosystems are remarkably resilient, with the ability to heal and regenerate even after significant disturbances. Whether through natural processes, the stabilizing influence of keystone species, or human intervention, ecosystems can find new life after destruction. However, the continued health of our planet's ecosystems depends on our ability to balance human activity with the preservation and restoration of the natural world. Conservation efforts that protect keystone species, restore habitats, and involve local communities are essential for ensuring that the cycles of healing and rebirth continue for generations to come.

Chapter 10: Finding Hope in Life's Darkest Moments

Life's path is often fraught with challenges and hardships that can sometimes feel insurmountable. Whether faced with personal loss, global crises, or the daily struggles of existence, it is in these darkest moments that the human spirit shines its brightest. Hope becomes a guiding light, leading individuals and communities toward recovery, growth, and renewal. This chapter explores the essential role that hope plays in our lives, the power of collective support, and the importance of finding meaning amidst adversity.

The Power of Hope: How Maintaining Hope Can Sustain Life Through the Darkest Times

Hope is a fundamental and deeply ingrained aspect of the human experience. It is the belief that, no matter how difficult the present may seem, there is a future worth striving for. When people find themselves in the depths of despair, whether due to illness, personal loss, or global calamities, hope serves as a beacon that offers the strength to continue moving forward.

Psychological Benefits of Hope

Finding the Light

Psychologically, hope plays a critical role in maintaining mental health and emotional resilience. Research has shown that people who maintain hope during challenging times are more likely to experience better overall well-being. This is because hope encourages positive thinking, motivates proactive behavior, and reduces feelings of helplessness.

- **Positive Outlook**: Hope helps shift focus away from the current hardship and onto the possibility of a better future. It reframes the narrative from one of defeat to one of potential recovery and success. In times of crisis, maintaining a hopeful mindset can keep individuals from sinking into despair.
- **Resilience and Perseverance**: Hope instills a sense of resilience, pushing people to persevere even when the odds are against them. For instance, individuals undergoing treatment for chronic illness often draw strength from the hope of recovery or at least an improvement in their quality of life. This sense of hope empowers them to endure difficult treatments and continue their battle.
- **Improved Problem-Solving Abilities**: Hopeful individuals tend to be more solution-focused. Instead of dwelling on the difficulties, they focus on finding ways to improve their situation. This optimism fuels creativity and resourcefulness, enabling people to

overcome obstacles they might have otherwise deemed impossible.

Historical and Personal Examples of Hope

Throughout history, there have been countless stories of individuals and groups who have triumphed over adversity because they held on to hope.

- **Victor Frankl and Holocaust Survival**: Victor Frankl, a psychiatrist and Holocaust survivor, famously wrote about his experiences in Nazi concentration camps. In his groundbreaking book, *Man's Search for Meaning*, Frankl emphasized that hope was crucial for survival in the camps. Those who retained a sense of purpose and a belief in a better future, no matter how distant, were more likely to endure the horrific conditions. Frankl's story illustrates how hope is not only a psychological asset but a literal lifeline in the direst of situations.
- **Nelson Mandela and the Struggle Against Apartheid**: Nelson Mandela's 27 years in prison, much of it in harsh conditions on Robben Island, is another example of how hope can sustain life through dark times. Mandela never lost sight of his belief that South Africa would one day be free from apartheid, and it was this hope that fueled his resilience. His optimism, even in the face of brutal

treatment, inspired others to continue the fight for justice.

- **Personal Journeys of Illness and Recovery**: On an individual level, stories of people who survive terminal illness often credit hope as a major factor in their recovery. Whether it's the hope of seeing their children grow up or experiencing life's joys again, hope acts as an emotional anchor that allows them to fight through pain and uncertainty.

The Role of Community: How Collective Effort Can Overcome Adversity

While hope is a powerful individual force, it is amplified when shared within a community. Humans are inherently social creatures, and we thrive on connection and support. During life's darkest moments, the collective effort of communities can be a crucial lifeline for individuals and the group as a whole. Communities provide emotional, physical, and mental support, enabling individuals to lean on each other and draw strength from shared experiences.

The Strength of Social Support

Numerous studies highlight the importance of social support in overcoming adversity. When individuals face hardship, whether it's economic, emotional, or physical, having a support network can dramatically improve their ability to cope.

- **Emotional Support**: Sharing one's fears, frustrations, and challenges with others can significantly reduce the burden of stress. The simple act of being heard and understood by others who may have gone through similar experiences fosters a sense of belonging and validation. During times of crisis, people find solace in knowing they are not alone in their struggle.
- **Practical Assistance**: Communities often come together to provide practical support during tough times. Whether it's delivering meals to someone recovering from surgery, offering financial help to those in economic distress, or organizing relief efforts during natural disasters, collective action can significantly ease the burdens individuals face.
- **Psychological Resilience**: The collective experience of hardship can also create a shared sense of purpose and resilience. People draw inspiration from each other, reinforcing the hope that together they can overcome adversity. In times of crisis, communities often strengthen their bonds, becoming more cohesive and resourceful.

Examples of Community Resilience

There are many instances in history where communities have come together to overcome significant challenges, reinforcing the idea that collective effort is essential in dark times.

- **Natural Disaster Recovery**: After natural disasters such as hurricanes, earthquakes, or wildfires, affected communities often band together to rebuild. In the aftermath of Hurricane Katrina in 2005, the people of New Orleans relied on each other for shelter, food, and emotional support. Volunteers from across the country joined the effort to help rebuild homes and provide care, demonstrating the power of collective action in the face of tragedy.
- **Global Crises**: During the COVID-19 pandemic, communities around the world witnessed an unprecedented level of solidarity. From frontline healthcare workers risking their lives to individuals organizing mutual aid groups, the pandemic showcased how community-driven support can help mitigate the impact of even the most global crises.
- **Civil Rights Movements**: Throughout history, civil rights movements have relied on community strength. The solidarity within the African American community during the Civil Rights Movement in the 1960s is one example. Individuals who faced violence, imprisonment, and discrimination found hope in their collective effort, knowing that together, they could change the course of history.

Finding Purpose: The Importance of Finding Light and Meaning During Difficult Journeys

In life's darkest moments, finding a sense of purpose can be the key to survival and eventual recovery. Purpose gives individuals something to strive for, a reason to endure the trials they face. Without purpose, it becomes easy to succumb to feelings of despair and hopelessness.

Purpose as a Guiding Light

Finding purpose doesn't necessarily mean achieving monumental goals; rather, it often involves discovering small but meaningful reasons to keep moving forward. For some, this might be caring for a loved one. For others, it might be the pursuit of a personal goal or the desire to make a positive impact on the world.

- **Victor Frankl's Concept of Purpose**: As mentioned earlier, Victor Frankl's philosophy emphasized that even in the most desperate situations, people could find meaning. Frankl believed that life's meaning was not something one simply stumbled upon but something each person must actively create. By focusing on a goal, task, or purpose, individuals can transcend their suffering and find the motivation to continue living.
- **Small Victories**: For those facing prolonged adversity, finding purpose often comes in small

steps. For a person recovering from a debilitating illness, their purpose might be as simple as regaining the ability to walk or spending time with loved ones. These seemingly minor goals can provide the motivation needed to persevere through difficulty.

- **Creativity and Contribution**: Many people find purpose by contributing to something larger than themselves. This might be through art, writing, activism, or helping others. During times of personal crisis, focusing on creative or altruistic activities provides a sense of fulfillment and purpose that transcends one's immediate suffering.

The Role of Spirituality and Philosophy

For many, purpose is closely tied to spiritual or philosophical beliefs. Faith traditions around the world emphasize finding meaning in suffering, whether through religious narratives or philosophical frameworks.

- **Religious Faith**: Many individuals draw strength from their faith, believing that their suffering has a greater purpose, whether it is part of a divine plan or a test of endurance. Religious communities also provide a supportive network where individuals can find solidarity and hope.
- **Stoicism and Acceptance**: Philosophies like Stoicism advocate for finding peace and meaning in

the acceptance of life's inherent struggles. The Stoic belief is that while we cannot control external events, we can control our response to them. This mindset encourages individuals to find purpose in their resilience, fortitude, and grace under pressure.

Conclusion: Life's Constant Search for Light

As this chapter has illustrated, life—whether in its most basic forms or through the complex experiences of human beings—always seeks the light, even in its darkest moments. Hope, community, and purpose form a trifecta of resilience that enables individuals and societies to weather the storm and emerge stronger on the other side. By cultivating hope, leaning on the support of others, and finding meaning in adversity, life proves time and time again that it can—and will—always find a way forward.

Conclusion: Life's Eternal Quest for the Light

Life's resilience is one of the most awe-inspiring phenomena in the natural world. From the simplest microorganisms to the most complex human societies, the unyielding capacity for survival, adaptation, and growth underpins the very fabric of existence. Whether it's the ability of bacteria to survive in the harsh vacuum of space, the tenacity of plants to sprout through cracks in concrete, or the sheer determination of human beings to overcome immense personal challenges, life demonstrates an incredible ability to push forward—an eternal quest for the light.

This book has delved into a wide variety of examples, each offering a unique perspective on how life finds a way to persist even when faced with overwhelming odds. From the early origins of life on Earth, through the process of evolution and adaptation, to the triumphs of human will and the recovery of ecosystems, every chapter has shown that life, in all its forms, possesses a remarkable drive to endure. This resilience is not just a physical phenomenon, but also a metaphor for the persistence of hope, renewal, and progress in the face of adversity.

Life's Tenacity in Extreme Conditions

Finding the Light

The capacity of life to thrive in environments previously thought to be uninhabitable is one of the most profound demonstrations of resilience. Whether it's the extremophiles that live in deep-sea hydrothermal vents, surviving without sunlight, or bacteria that endure the radiation-drenched conditions of outer space, these life forms redefine the limits of what we consider habitable. They show us that life is not fragile—it's adaptable. This adaptability allows it to take root in even the most seemingly inhospitable environments, demonstrating that survival is not limited to Earth's temperate conditions but may be a universal trait wherever favorable conditions arise.

Moreover, these extreme environments often serve as laboratories for understanding the future of life on Earth and the possibility of life beyond it. The discovery of microbes living in the Antarctic ice or deep within Earth's crust forces us to rethink life's boundaries, challenging previous scientific assumptions. It opens our minds to the idea that life is far more robust than once imagined, able to harness energy from sulfuric compounds, withstand boiling temperatures, or exist in complete darkness. This suggests that wherever conditions arise that allow energy to flow—however scarce—life may find a way to flourish.

Evolution: Nature's Blueprint for Survival

Evolution is another key mechanism that underpins the persistence of life. Through the process of natural selection, life forms continuously adapt to their environments, passing on advantageous traits to future generations. Evolution is life's blueprint for survival in a constantly changing world, a testament to the resilience of species that have evolved over millions of years to meet new challenges. It has allowed life to diversify, filling every ecological niche, from the highest mountains to the deepest oceans, from the hottest deserts to the coldest tundra.

In this process, we observe the dynamic relationship between challenge and adaptation. Species that once seemed vulnerable to extinction find ways to thrive by adapting their behaviors, physiology, or reproductive strategies. For instance, animals like polar bears have adapted to survive in the frigid Arctic, while desert plants like cacti have developed mechanisms to store water during prolonged droughts. Every corner of the Earth is home to species that have uniquely adapted to the challenges of their environment. This ongoing process ensures that life remains resilient, constantly changing and evolving to meet new environmental pressures.

The Human Spirit: A Pinnacle of Resilience

Finding the Light

Perhaps the most poignant example of life's resilience is found in the human spirit. Throughout history, individuals and societies have faced and overcome tremendous adversity—wars, natural disasters, disease, and personal tragedies. Yet, in every instance, humanity's innate drive to survive, rebuild, and thrive has been evident. Whether through innovation, collaboration, or sheer force of will, humans have consistently shown an ability to rise above even the most challenging circumstances.

The triumph of the human spirit is not just a story of survival but one of growth and transformation. People who have faced severe challenges often emerge stronger, more resilient, and more compassionate. The capacity to find hope, even in the darkest of times, is one of the most powerful aspects of human resilience. It's what drives communities to rebuild after disasters, individuals to persevere through personal struggles, and societies to push for progress despite overwhelming odds.

Moreover, humanity's ability to adapt to and overcome adversity is increasingly seen in the face of global challenges, such as climate change, pandemics, and social inequality. As we grapple with these issues, we are reminded that life—human life in particular—has the capacity not just to survive but to innovate and create solutions that ensure future resilience. Through science, technology, and collective action, we are constantly finding

Finding the Light

new ways to adapt to the changing world, ensuring that life continues to flourish.

Nature's Recovery: Life Rebounds

The natural world, too, offers countless examples of resilience and renewal. Even after the most devastating of natural disasters—wildfires, tsunamis, hurricanes, or volcanic eruptions—nature always finds a way to recover. Forests regrow after fires, coral reefs regenerate after bleaching events, and ecosystems heal after pollution or deforestation. These processes of regeneration remind us that, while life may be disrupted, it is rarely defeated.

This ability to rebound after devastation is a powerful metaphor for the human experience. Just as ecosystems eventually heal and return to equilibrium, individuals and societies, too, have the capacity for renewal. The regrowth of forests after a fire, for instance, not only symbolizes the cycle of destruction and rebirth but also demonstrates how new life can emerge from the ashes, often stronger and more resilient than before. This regenerative power of life, seen so clearly in nature, reflects the inherent drive to find the light after even the darkest of times.

Life's Universal Drive

Ultimately, the resilience of life is a universal drive that transcends species, environments, and circumstances.

Finding the Light

Whether it is the smallest microbe or the largest mammal, life's pursuit of survival and adaptation is relentless. The quest for light—both literally and metaphorically—drives the evolutionary processes that sustain all living things. It is what compels plants to grow towards the sun, what leads animals to migrate vast distances for survival, and what fuels humanity's efforts to overcome challenges and improve the world.

At its core, this drive reflects the fundamental truth that life is not passive; it actively seeks to grow, evolve, and push through the barriers placed before it. It is a force of nature, propelled by the desire not only to exist but to thrive. This capacity for persistence in the face of adversity is what makes life so remarkable and awe-inspiring.

Conclusion: A Lesson in Hope and Renewal

As we conclude this exploration of life's resilience, we are reminded that the quest for light is not limited to the natural world. It is also deeply embedded in the human experience. Just as ecosystems recover from disaster and species adapt to survive, we too, as individuals and as a collective, have the capacity for renewal, growth, and transformation.

Life's eternal quest for the light teaches us that no matter how dire the circumstances, there is always hope. There is always a way forward, a path to growth, and a chance for new beginnings. It is this unyielding resilience that ensures that life, in all its forms, will continue to find a way, no matter the challenges it faces. The light, as elusive as it may sometimes seem, is always within reach for those who persevere.

And so, as we look to the future, both for ourselves and for the world around us, we can take comfort in the knowledge that life will continue to endure, adapt, and thrive. Life always finds a way—because that is its nature, its driving force, and its greatest strength.

Bibliography

1. **The Selfish Gene** by Richard Dawkins

Explores the theory of evolution and natural selection, offering insight into how life adapts and thrives despite adversity.

2. **The Rise of Life on Earth** by Joyce Carol Oates

Delves into the early history of life and how biological systems evolved to survive and thrive in changing environments.

3. **The Immortal Life of Henrietta Lacks** by Rebecca Skloot

A fascinating story about human cells that continue to live and grow decades after a person's death, showing the tenacity of life at the cellular level.

4. **Man's Search for Meaning** by Viktor E. Frankl

An exploration of human resilience and survival in the face of unimaginable suffering during the Holocaust, focusing on finding hope in life's darkest moments.

5. **The Sixth Extinction: An Unnatural History** by Elizabeth Kolbert

Finding the Light

A look at the five mass extinctions in Earth's history and the resilience of life in the aftermath, with a focus on the current anthropogenic crisis.

6. **Endurance: Shackleton's Incredible Voyage** by Alfred Lansing

The legendary true story of Ernest Shackleton's Antarctic expedition and the ultimate display of human resilience and survival against impossible odds.

7. **The Hidden Life of Trees: What They Feel, How They Communicate – Discoveries from a Secret World** by Peter Wohlleben

This book reveals the incredible ways in which trees and forests adapt to survive, thrive, and communicate, showing nature's resilience.

8. **Sapiens: A Brief History of Humankind** by Yuval Noah Harari

Chronicles the journey of human beings from the dawn of Homo sapiens to today, focusing on our adaptability and survival instincts.

9. **Survival of the Sickest: The Surprising Connections Between Disease and Longevity** by Sharon Moalem

Offers a fascinating perspective on how certain genetic traits and diseases have helped humans survive and evolve.

10. **Life in the Extremes: The Science of Survival** by Frances Ashcroft

Explores how life thrives in the most extreme environments on Earth, from deep ocean trenches to high mountain peaks.

11. **Why Zebras Don't Get Ulcers** by Robert M. Sapolsky

A scientific exploration of stress, survival, and the human body's remarkable ability to adapt and find equilibrium in tough situations.

12. **The Botany of Desire: A Plant's-Eye View of the World** by Michael Pollan

This book takes a closer look at how plants have evolved to thrive by adapting to human desires and needs, showcasing the ingenuity of life.

Acknowledgments

Writing *Finding the Light: How Life Always Finds a Way* has been an incredible journey, and I am deeply grateful to everyone who has helped me along the way.

First and foremost, I would like to express my profound gratitude to my family and friends for their unwavering support and encouragement. Your belief in me kept the light of inspiration burning bright throughout this process. Your patience and understanding during the long hours of writing and research have meant the world to me.

A heartfelt thanks goes to my editor, whose keen insights and meticulous attention to detail greatly improved this book. Your guidance shaped the final version of this work, and I am thankful for your expertise and passion.

To the researchers, scientists, and explorers who continue to uncover the mysteries of life's resilience, I owe a deep debt of gratitude. Your tireless efforts to push the boundaries of human knowledge inspired many of the ideas within these pages.

I would also like to acknowledge the countless authors and thought leaders whose work on survival, adaptation, and the tenacity of life has enriched my understanding of the topic. Your contributions to this field have shaped my perspectives and deepened my appreciation for life's unyielding nature.

Finally, I want to extend my appreciation to every reader who embarks on this journey with me. Your curiosity and desire to explore the wonders of life's resilience are what make this book possible. It is my hope that *Finding the Light* brings inspiration, hope, and a deeper understanding of life's ability to endure.

Thank you all for being part of this journey.

Sincerely,

Zahid Ameer
Versatile Indie Author/Ghostwriter

Disclaimer:

The content of this book is for informational and educational purposes only. While every effort has been made to ensure the accuracy and reliability of the information presented, the author and publisher make no representations or warranties of any kind, express or implied, regarding the completeness, accuracy, reliability, suitability, or availability of the content.

This book includes scientific discussions, historical events, and personal anecdotes, which are provided as general information. It is not intended as professional advice, nor should it be used as a substitute for consulting with experts in relevant fields such as biology, ecology, psychology, or medicine.

The author and publisher disclaim any liability for any loss or damage arising directly or indirectly from the use of or reliance on this book. Readers are encouraged to seek professional advice before making any decisions or taking action based on the content of this book.

About me

I am Zahid Ameer, hailing from the vibrant country of India. As an author, ghostwriter, bibliophile, online affiliate marketer, blogger, YouTuber, graphic designer, and animal lover, I have woven my passions into a unique tapestry that defines my life's work.

Born and raised in India, I have always possessed a deep love for literature. With an insatiable appetite for books, I have amassed an impressive collection of around 1,500 titles, predominantly in English. My passion for reading brings me immense joy and serves as a source of inspiration for my writing endeavors.

I have compiled an impressive portfolio of written works as an author and ghostwriter. With a captivating writing style and an innate ability to craft engaging narratives, I bring my stories to life, captivating readers from all walks of life. My wide range of interests and experiences contribute to the richness of my writing, allowing me to connect with my audience on a heartfelt level effortlessly.

Beyond my literary pursuits, I have also established a strong presence on various digital platforms. I utilize my YouTube channel and blog to raise awareness about all types of knowledge and to share heartwarming stories of animals. Using my platform to shed light on important issues, I strive to create a world where humans and animals can coexist harmoniously.

In addition to my work as an author, I have also dabbled in the world of affiliate marketing. With my webpreneur spirit, I have ventured into online marketing, leveraging my knowledge and skills to promote products and services that align with my values.

However, my most cherished role is that of a father. Family is at the core of my being, and everything I do is centered around creating a better future for my loved ones. My dedication to my family is evident in my passion for personal growth and my relentless pursuit of success. Through my various endeavors, I strive to set an example of perseverance and ambition for my children, inspiring them to chase their dreams unapologetically.

In a world where specialization often dominates, I defy convention by embracing multiple passions and excelling in diverse fields. My love for books, animals, and family has become the driving force behind my achievements. By the grace of Almighty God, my unique blend of characteristics has allowed me to leave an indelible mark on the world, enriching the lives of those I encounter along the way.

To your grand success in life,

Zahid Ameer
Versatile Indie Author/Ghostwriter

www.ingramcontent.com/pod-product-compliance
Lightning Source LLC
Chambersburg PA
CBHW071059240526
45471CB00016B/2161